MR. WALLACE ON THE PHENOMENA OF VARIATION AND GEOGRAPHICAL DISTRIBUTION AS ILLUSTRATED BY THE MALAYAN PAPILIONIDÆ

(1864)

BY

ALFRED RUSSEL WALLACE

British Library Cataloguing-in-Publication Data
A catalogue record for this book is available from the
British Library

Alfred Russel Wallace

Alfred Russel Wallace was born on 8th January 1823 in the village of Llanbadoc, in Monmouthshire, Wales.

At the age of five, Wallace's family moved to Hertford where he later enrolled at Hertford Grammar School. He was educated there until financial difficulties forced his family to withdraw him in 1836. He then boarded with his older brother John before becoming an apprentice to his eldest brother, William, a surveyor. He worked for William for six years until the business declined due to difficult economic conditions.

After a brief period of unemployment, he was hired as a master at the Collegiate School in Leicester to teach drawing, map-making, and surveying. During this time he met the entomologist Henry Bates who inspired Wallace to begin collecting insects. He and bates continued exchanging letters after Wallace left teaching to pursue his surveying career. They corresponded on prominent works of the time such as Charles Darwin's *The Voyage of the Beagle* (1839) and Robert Chamber's *Vestiges of the Natural History of Creation* (1844).

Wallace was inspired by the travelling naturalists of the day and decided to begin his exploration career collecting specimens in the Amazon rainforest. He explored the Rio Negra for four years, making notes on the peoples and languages he encountered as well as the geography, flora, and fauna. On his return voyage his ship, Helen, caught fire and he and the crew were stranded for ten days before being picked up by the Jordeson, a brig travelling from Cuba to London. All of his specimens aboard Helen had been lost.

After a brief stay in England he embarked on a journey to the Malay Archipelago (now Singapore, Malaysia, and Indonesia). During this eight year period he collected more than 126,000 specimens, several thousand of which represented new species to science. While travelling, Wallace refined his thoughts about evolution and in 1858 he outlined his theory of natural selection in an article he sent to Charles Darwin. This was published in the same year along with Darwin's own theory. Wallace eventually published an account of his travels *The Malay Archipelago* in 1869, and it became one of the most popular books of scientific exploration in the 19th century.

Upon his return to England, in 1862, Wallace became a staunch defender of Darwin's landmark work *On the Origin of Species* (1859). He wrote responses to those critical of the theory of natural selection, including 'Remarks on the Rev. S. Haughton's Paper on the Bee's Cell, And on the Origin of Species' (1863) and 'Creation by Law' (1867). The former of these was particularly pleasing to Darwin. Wallace also published important papers such as 'The Origin of Human Races and the Antiquity of Man Deduced from the Theory of 'Natural Selection" (1864) and books, including the much cited *Darwinism* (1889).

Wallace made a huge contribution to the natural sciences and he will continue to be remembered as one of the key figures in the development of evolutionary theory.

Wallace died on 7th November 1913 at the age of 90. He is buried in a small cemetery at Broadstone, Dorset, England.

MR. WALLACE ON THE PHENOMENA OF VARIATION AND GEOGRAPHICAL DISTRIBUTION AS ILLUSTRATED BY THE MALAYAN PAPILIONIDÆ

Different groups of animals have very unequal values as illustrative of variation and distribution, some being not sufficiently rich in species, others having too limited a range, while many to which these objections do not apply are yet too imperfectly known to furnish us with data sufficiently detailed and accurate. The diurnal Lepidoptera, or butterflies, however, seem to have all the necessary qualities. They are very numerous, and their extreme beauty has led to their having been assiduously collected in all parts of the world. Their immensely developed wings are covered with scales which imitate the rich hues and delicate surfaces of satin or velvet, glitter with metallic lustre, or glow with the varying tints of opal; these colours are disposed in an endless variety of patterns, and the gaily painted surface acts as a register of the minutest changes of organization, and exhibits on an enlarged scale the effects of the climatal and organic conditions which have influenced more or less profoundly the organization of every living thing.

The Papilionidæ are one of the most important families of butterflies. They are very highly organized, and present differences both in the larva and imago state of a very striking character. They are also among the largest, the most majestic, and the most elegant of butterflies, and, though pretty generally distributed over the whole earth, are especially abundant in the tropics, where they attain their maximum of size and beauty. South America, North India, and the Malay Islands are the regions where these fine insects occur in the greatest profusion, and where they actually become a not unimportant feature in the scenery; for the stately flight and gorgeous colouring of the larger ones, which are seven or eight inches across the wings, render them even more conspicuous than the majority of birds.

No less than 120 species of the family inhabit the Malay Archipelago, distributed throughout the islands in varying proportions, Borneo containing the largest number (twenty-nine species), while many of the smaller islands have only from six to ten species. A careful study of large series of these insects, collected in about thirty islands from Sumatra to New Guinea, has brought to light a number of curious facts in variation and geographical distribution which it was the chief object of the paper to illustrate. The author considered that variation was by no means one simple fact, but included distinct phenomena which have been very often confounded. He would class the phenomena of variation under the heads of--1st, Simple Variability; 2nd, Dimorphism, or Polymorphism; 3rd, Local Forms; 4th, Co-existing Varieties; 5th, Races, or Sub-species; and, 6th, True Species.

Simple variability includes all cases of great instability of specific form, or those in which the offspring differ more or less from the parents, but differ irregularly, and, as it were, accidentally. In such cases, so great is the tendency to vary that it is difficult to find two individuals exactly alike. One of the Malayan species (*Papilio severus*) exhibits this in a remarkable degree, and several others less strikingly. It may be called irregular variability, and is of the same nature as that so characteristic of domestic breeds.

Polymorphism, or dimorphism, differs from simple variability in this--that the offspring differ from the parents in a considerable degree, and in a manner more or less constant and regular; so that, of the offspring of a single pair, some will resemble their parents, while others will differ from them; but the differences will be tolerably fixed and definite, and intermediate varieties will never occur. A large and handsome Malayan butterfly, *Papilio memnon*, is a good case of dimorphism. The male is nearly uniform bluish black, with rounded hind wings, and never varies. One portion of the females resemble the males in shape, but are coloured brown or ashy, and with more or less white markings on the hind wings. Another set of females are found, however, which differ remarkably in the shape of the wings, the under ones being lengthened behind into a large spoon-shaped tail; and they have also white lines radiating from the base of the wing. Intermediates in form or colour between these two kinds of females never occur. These very distinct kinds of female insect do not produce young like themselves only; for, from eggs laid by one of them, both kinds of butterflies are produced, as well as the male, which is different from either. It is just as if a blue-eyed flaxen-haired white man had two wives, one a straight-haired red-skinned Indian squaw, the other a woolly-headed coal-black negress; and all the boys from either mother were real white boys just like their father, while all the girls were either pure negro or pure Indian; but the Indian mother should sometimes have negro daughters, and the negro mother Indian daughters. Such a thing seems absurd and contrary to nature; yet this is exactly what takes place in the butterfly called *Papilio memnon*-- the males are always exactly like the male parent, while the female offspring of each mother are partly like and partly unlike herself.

The butterfly called *Papilio pammon*, inhabiting all parts of India, is another case. It is nearly black, with a band of white spots along the margin of the fore wings and across the middle of the hind wings. Some of the females are exactly like the male, having only a small additional red spot behind. The most abundant females are, however, quite different, having a large white and brick-red patch on the hind wings and a row of red spots. This was long supposed to be a different species named *Papilio polytes*; but, from the eggs laid by it, *Papilio pammon* was produced as well as others resembling itself; and a further proof is that no male of *P. polytes* has ever yet been found, although the species is very common. There is, however, in India yet another butterfly, named *Papilio romulus*, of which no male has ever been found, although female specimens exist in every collection; and, from a careful examination of this insect, the author was convinced that it was really a third form of the female of *Papilio pammon*. To parallel this case, we must suppose that our white man has a third wife, white like himself, and that the children of both white, black, and red wives, if boys, are all alike white; but, if girls, may be white, black, or red, and often without any regard to the colour of their mothers. Further east, in the Moluccas and New Guinea, there is a butterfly named *Papilio ormenus*, which has also three kinds of females, but in this case all are different from the male; just as if our supposed polygamist had an oblique-eyed yellow Chinese wife in addition to his black and red partners. The more exact parallel would be, however, to suppose an island inhabited only by white men, with black, red, and yellow women, and that, after many generations had passed away, the men all remained pure white, and the women of their respective colours and races equally pure. This would exactly represent the phenomena of *polymorphism*, which is the ordinary course of nature in several species of insects in various parts of the world, and will sufficiently show that it is quite a different thing from simple variation, with which it has hitherto been confounded.

There are, however, other species which show how this state of things may have been brought about. One inhabiting the Philippine Islands (*Papilio alphenor*) has females which vary very much while the males are very different and constant. Now Mr. Darwin has shown that very small differences in external appearance are often accompanied by a very important difference in constitution, so that one variety, being better adapted to surrounding circumstances than another, may increase and flourish, while its companion diminishes, and ultimately dies out. But, in the case of an insect producing a very variable female, it may come to pass that the extreme varieties will each possess some qualities which render them better adapted to exist than the intermediate forms, which will therefore die out and leave only the extremes. The difficulty remains, however, of accounting for the fact that these intermediate varieties are not continually reproduced; and this can only be explained by the tendency of natural selection being to preserve continually more and more of the offspring of those individuals which produce only one or both of the favourable forms, and few or none of the unfavourable ones; and also from the fact that among the offspring of those which produce the largest proportion of the extreme favourable varieties will be found those most perfect and highly-developed individuals which, when adverse circumstances threaten the extermination of the whole race, will alone maintain their existence and transmit to the whole succeeding race the physiological as well as external peculiarities of their parents.

The cases of different broods of the same species having very distinct characters, as in many European butterflies, was mentioned as a somewhat different though allied phenomenon, which may be called seasonal or alternate dimorphism. Ants exhibit another case in three or four differently-formed sets of individuals being produced from the same parents; but here each form was of use to the whole community, and Mr. Darwin has shown how they could have been produced by "natural selection;" this might be called structural or economical dimorphism. Yet another is that shown to exist in many plants (as *Linum* and *Primula*) by Mr. Darwin, where the reproductive organs are modified so that each form bears a definite functional relation to the other. This may be called sexual or reproductive dimorphism. All these different cases, however, have this character in common, that the union of the distinct forms does not produce intermediate varieties, but reproduces the two or more parent forms altogether pure; whereas, when simple varieties or races intermingle, the offspring seldom resemble either parent exactly, but are more or less intermediate between them.

Passing over the subject of local forms, races, and species, which were treated of at some length, we come to that of "Variation as specially influenced by locality," on which the author announced that he had some new and interesting facts to communicate. On carefully comparing the allied species of butterflies from different districts of the archipelago and from separate islands, it was found that a special character was in some cases communicated to the majority of the species in several families. In the Papilionidæ the following were the results:--

1st. The species of the Indian region (Sumatra, Java, &c.) are almost invariably smaller than the allied species of Celebes and the Moluccas.

2nd. The species of New Guinea and Australia are also, though in a less degree, smaller than the nearest species or varieties of the Moluccas.

3rd. In the Moluccas the species and individuals inhabiting the island of Amboyna are the largest.

4th. The species of Celebes equal, and sometimes surpass, in size those of Amboyna.

5th. The species and varieties of Celebes possess a striking character in the form of the anterior wings different from that of the allied species and varieties of all the surrounding islands.

6th. Tailed species in India and the western islands lose their tails as they spread eastward through the archipelago.

Details were given and diagrams, as well as specimens exhibited showing the peculiarities of size and form above recapitulated. The most interesting case, however, was that of the island of Celebes, almost all the Papilionidæ, Pieridæ, and some of the Nymphalidæ of which had acquired a peculiar curve of the upper wings, amounting in some instances to an abrupt bend. The accompanying figure shows the outline of the upper wing in *Papilio gigon* and *Papilio demolion*, very closely allied species, the former (upper figure) inhabiting Celebes only, the latter (lower figure) common in Borneo, Java, and Sumatra.

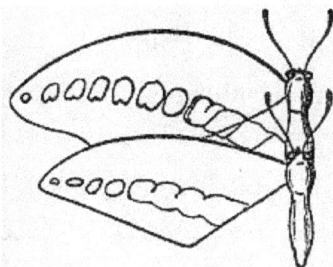

In no less than thirteen species of Papilio, ten Pieridæ, and five or six Nymphalidæ, the difference of form between the species or varieties peculiar to Celebes, and those most closely allied which inhabit all the other parts of the archipelago, is of exactly the same character and of nearly equal amount. The difference is the same whether we compare the small species of Java or the large ones of the Moluccas with those of Celebes, showing that the causes which have produced it are distinct from those which have led to increase of size. The difference is equally perceptible both in the *varieties* and in the *species* peculiar to Celebes; and this is held to be in favour of the doctrine that species and varieties are really of the same nature, and differ only in degree, since the local causes which have been at work in Celebes have acted on both in a manner perfectly identical.

In attempting to explain the possible origin of this curious phenomenon, it was pointed out that one species (*Papilio polyphontes* Bd.), inhabiting Celebes, did not offer any differences of form when compared with its nearest allies in the surrounding islands; and that this unchanged species formed part of a group which, from being slow of flight, abundant in individuals, and especially from being the object of mimicry by other groups, was considered to have some special and hidden protection independent of flight. If, therefore, the butterflies of Celebes acquired these longer and more curved wings, owing to the persecution of bird or insect enemies from whom they could only escape by increased powers of flight, it is evident that those which had already some other means of protection would receive no benefit from a change in the form of their wings, and therefore could not acquire it by the action of "natural selection." This also explains why none of the Danaidæ are so modified, for they are universally the objects of mimicry by other groups, and are therefore already protected. The large thick-bodied Nymphalidæ also fly very rapidly by the power of their muscles rather than the length of their wings, and are probably never captured on the wing; they have, therefore, not received any similar modification, because they did not require it.

In like manner the weak, obscure, brown Satyridæ and the small, active Lycenidæ and Hesperidæ, all of which have unmodified wings, secure themselves from attack rather by concealment and peculiar habits than by direct power of flight. Although, therefore, we are not able to point out the peculiar enemies to Lepidoptera which have existed in Celebes alone, we may be sure that the singular alteration in the form of wing in so many of the butterflies of that island, as well as the changes of form and size induced in other parts of the archipelago, are the effects of that complicated action and reaction of all living things upon each other in the struggle for existence, which continually tends to readjust disturbed relations, and to bring every species into harmony with the ever-varying conditions of the surrounding universe.

The subject of mimicry was then explained, and it was shown by illustrative specimens that a number of the Malayan and Indian *Papilios* resembled very closely species of Danaidæ inhabiting the same districts. These Danaidæ are almost a nuisance to the collecting entomologist from their abundance and ubiquity. Every garden, every roadside, the suburbs of every village are full of them--indicating very clearly that their existence is an easy one, and that they are free from persecution by the foes that keep down the population of less favoured races. Their strong and peculiar odour is believed to be the cause of their safety, and they are for this reason habitually passed over by insectivorous creatures. When, therefore, another insect which has not this peculiar scent resembles one of the Danaidæ of the same district, it also is passed over by mistake, and thus gains an advantage. When this resemblance is general and slight, the benefit will be correspondingly small. Such a slight general resemblance may, however, evidently occur accidentally between very differently constructed animals; and what is important to observe is, that when once this happens the resemblance and the corresponding advantage will necessarily go on increasing by the action of "natural selection." For example, let a *Papilio* resemble a *Danais* so slightly that it is only mistaken for it by very unobservant birds, &c., or at a considerable distance. Even this will be some advantage to it, for many individuals that would otherwise have been devoured will now live and leave offspring. But every animal varies more or less, sooner or later. Among the varieties of this *Papilio* some must be more like, some less like the *Danais*. The former will escape persecution more than the latter--will increase, therefore, while the latter will diminish. In each succeeding generation this preservation of the more like and the destruction of the less like will go on, which must slowly, but surely, produce a gradually increasing likeness, till the one insect can hardly be distinguished from the other. Such cases occur everywhere. Some of the more remarkable have been pointed out by Mr. Bates as occurring in South America, where every streak and spot, as well as the exact outline, has been copied by insects of a quite different structure. In tropical Africa cases quite as extraordinary occur, two or more insects of widely different structure coming outwardly to resemble a third equally different from both; and, in every case whether there are two or three, one of them is a *Danais*, or belongs to one or two other groups which have also some special protection. In every case, too, the insects which resemble each other inhabit the same country, and, in cases where their habits are known, are found always to frequent the same places. This takes the phenomenon altogether out of the region of accidental resemblance, for it is evident that the chances of one insect resembling another are much greater when we compare those of all parts of the world than when we confine ourselves to those inhabiting the same country, the same locality, and even the same station. Yet cases of this very close resemblance between species of distinct families and genera are very rare in the former case--just as rare, in fact, as accidental resemblances should be; whereas, in the latter case, where, according to the law of chances, they should be infinitely more rare, they are so frequently to be met with in every part of the world that they may almost be called common. There must be, therefore, in the cases pointed out by Mr. Bates and Mr. Wallace, a true relation between the species so resembling each other; and, if the explanation given by them be objected to, it should be shown that it is inadequate to account for the observed facts. Instead of this, however, the objection is made that there ought to be more cases of mimicry if this explanation is true; that almost all *Pieridæ* should imitate *Heliconias*, and almost all *Papilios* should have come to resemble *Danaidæ*. To this the answer appears sufficient, that it is only when, among the countless variations of insects, a remote resemblance accidentally occurs between the distinct groups, that "natural selection" can begin to act to increase that resemblance for the creature's benefit. The remaining species of the group will be preserved in a different manner. Some will acquire increased powers of flight, others facilities for concealment, others, again, increased reproductive powers, which will enable them to resist the persecution of their enemies. And it will depend, at some critical period of the existence of the species, on the proportionate amount and rapidity of variation in these different directions to determine the special means by which its perpetuation shall be secured. The comparative isolation of cases of mimicry is therefore no argument either against its serving as a means of protection, or against its having been produced in the manner here indicated. No less than fourteen Indian and Malayan Papilios mimic species of distinct groups, in several cases so closely that they have been placed by entomologists in the same species, although really having no close affinity; and the writer was constantly deceived by them on the wing. We cannot wonder, therefore, that birds and insects also confound them.

The arrangement and geographical distribution of the Papilionidæ formed the concluding section of the paper. The family is very richly represented in the Malay islands, more than a quarter of all the known species being found there. In Africa about 40 species of Papilio exist, in Tropical Asia 65, in South America 120, or about the same number as in the Malay Archipelago. The area of the two countries is, however, very different; for, while South America contains more than 5,000,000 square miles, a line encircling the whole of the Malay islands would only include an area of 2,700,000 square miles, of which the actual land surface would not exceed 1,000,000 square miles. This comparative richness was shown to be partly due to the breaking up of the area into numerous isolated tracts, which led to the separation of many forms which, on a continental area, would probably exist as widely diffused and variable species. Dividing the Malayan Papilionidæ into twenty groups of the most closely allied species, it was shown that seven of those groups were confined to the Indo-Malayan and three to the Austro-Malayan region, indicating the same division of the archipelago between the Indian and Australian regions which has already been established from a consideration of the distribution of Mammalia and Birds. The degree of the relationship of the islands with each other was also indicated by the relative proportion of the species common to them, and was often opposed to that indicated by their geographical position or their physical characteristics. A striking example was that of Java and Sumatra, which are so closely connected by intervening islands in the Straits of Sunda, and have such a similarity in the continuous chain of volcanoes which passes through them, that we can hardly avoid the idea of their recent separation. This idea, however, is erroneous, for they have really less common resemblance in their natural productions than either has with Borneo--separated from both by a wide extent of sea, and differing from both in its non-volcanic structure. The relations of the Papilionidæ of these islands are as follow:--

Sumatra... 21 species} 20 species common to both islands.
Borneo . . . 29 species

Java . . . 27 species } 20 species common to both islands.
Borneo . . . 29 species

Sumatra...21 species } 11 species common to both islands.
Java . . . 27 species

Showing that both Sumatra and Java have a much closer relationship to Borneo than they have to each other.

This exactly confirms a similar deduction from the Birds and Mammalia of the three islands, and may therefore be held to prove that the only recent connexion of Sumatra and Java has been through Borneo.

Again, Borneo and Java have each only *two* species quite peculiar to them, and Sumatra not *one*. Celebes, however, not separated from them geographically more than they are from each other, has *seventeen* species altogether restricted to it. Further east, no island has more than *five* species confined to it. Celebes, therefore, stands alone, and has an individuality comparable with that of extensive groups rather than with single islands; and, though situated in the very midst of the archipelago, and surrounded on every side by islets which seem to afford the greatest facilities for the migration and intercommunication of the productions of it and the rest of the archipelago, yet retains a character of its own.

The great peculiarities of Celebes in its mammals, birds, and insects were then recapitulated, as well as the singular modifications of form in its Lepidoptera, which stamps the most dissimilar species with a mark distinctive of their common birthplace. It was argued that such phenomena could not be explained by the simple doctrine of special creations. There were so many signs of gradual modification and dependence upon physical and organic changes that we cannot believe it to be all a delusive appearance, any more than we can believe that strata were never deposited in primæval oceans, or that the fossils collected by the geologist are no true record of a former living world, but were all created just as they now appear.

All the curious phenomena here brought forward were believed by the author to be immediately dependent on the last series of changes, organic and inorganic, in these regions; and, as the phenomena presented by the island of Celebes differ from those of all the surrounding islands, it could, he conceived, only be because the past history of Celebes has been to some extent unique and different from theirs. More evidence was wanted to determine in what that difference consisted. At present one deduction only could be made-- viz., that Celebes represented one of the most ancient parts of the archipelago; that it has been formerly more completely isolated both from India and from Australia than it now is; and that, amid all the mutations it has undergone, a relic of the fauna and flora of some more ancient land has been here preserved to us. The full peculiarity and interest of the country had only been understood since Mr. Wallace had been able to compare the productions of Celebes side by side with those of the rest of the archipelago; and he concluded by expressing a hope that some enterprising naturalist might devote himself to its more detailed examination, since he was sure that no single island on the globe promised so well to repay a careful research into its past and present history.

www.ingramcontent.com/pod-product-compliance
Lightning Source LLC
Chambersburg PA
CBHW021340290326
41933CB00038B/1000